Schomburg Plaza Fire
New York City (Harlem)

Investigated by: Philip Schaenman

This is Report 004 of the Major Fires Investigation Project conducted by TriData Corporation under contract EMW-86-C-2277 to the United States Fire Administration, Federal Emergency Management Agency.

Homeland Security

Department of Homeland Security
United States Fire Administration
National Fire Data Center

U.S. Fire Administration Fire Investigations Program

The U.S. Fire Administration develops reports on selected major fires throughout the country. The fires usually involve multiple deaths or a large loss of property. But the primary criterion for deciding to do a report is whether it will result in significant "lessons learned." In some cases these lessons bring to light new knowledge about fire--the effect of building construction or contents, human behavior in fire, etc. In other cases, the lessons are not new but are serious enough to highlight once again, with yet another fire tragedy report. In some cases, special reports are developed to discuss events, drills, or new technologies which are of interest to the fire service.

The reports are sent to fire magazines and are distributed at National and Regional fire meetings. The International Association of Fire Chiefs assists the USFA in disseminating the findings throughout the fire service. On a continuing basis the reports are available on request from the USFA; announcements of their availability are published widely in fire journals and newsletters.

This body of work provides detailed information on the nature of the fire problem for policymakers who must decide on allocations of resources between fire and other pressing problems, and within the fire service to improve codes and code enforcement, training, public fire education, building technology, and other related areas.

The Fire Administration, which has no regulatory authority, sends an experienced fire investigator into a community after a major incident only after having conferred with the local fire authorities to insure that the assistance and presence of the USFA would be supportive and would in no way interfere with any review of the incident they are themselves conducting. The intent is not to arrive during the event or even immediately after, but rather after the dust settles, so that a complete and objective review of all the important aspects of the incident can be made. Local authorities review the USFA's report while it is in draft. The USFA investigator or team is available to local authorities should they wish to request technical assistance for their own investigation.

For additional copies of this report write to the U.S. Fire Administration, 16825 South Seton Avenue, Emmitsburg, Maryland 21727. The report is available on the Administration's Web site at http:// www.usfa.dhs.gov/

U.S. Fire Administration

Mission Statement

As an entity of the Department of Homeland Security, the mission of the USFA is to reduce life and economic losses due to fire and related emergencies, through leadership, advocacy, coordination, and support. We serve the Nation independently, in coordination with other Federal agencies, and in partnership with fire protection and emergency service communities. With a commitment to excellence, we provide public education, training, technology, and data initiatives.

 Homeland
Security

TABLE OF CONTENTS

EXECUTIVE SUMMARY 1

SUMMARY OF KEY ISSUES 2

INTRODUCTION 2

LESSONS LEARNED 3

    Code Enforcement 3

    Fireground Perceptions and Communication 3

    Fire Dispatchers 4

    After the Fire: Department Critiques and Employee Assistance 4

    Public Education 4

ATTACHMENTS 7

    1. Fire Department of the City of New York Board of Inquiry report on the fire. 9

    2. FDNY press release on the Board of Inquiry's findings – July 27, 1987. 16

    3. Photographs. 19

Schomburg Plaza Fire
New York City (Harlem)
March 22, 1987

EXECUTIVE SUMMARY

A fire originating in the compactor chute of a 35-story high-rise apartment building in the Harlem area of New York City caused the deaths of seven building residents. Several code enforcement and fire department operational problems may have contributed to the loss.

The U.S. Fire Administration had planned to investigate this fire because of its many important lessons but would only do so with the express permission of appropriate authority. The Fire Department of New York (FDNY) requested that the investigation be delayed until a preliminary internal investigation was completed.

When the preliminary FDNY report was issued, the Fire Administration found it to be of such high quality and candor that an additional investigation did not seem likely to add much to the lessons of interest nationally.

This report summarizes some of the lessons learned on the FDNY report and discussions with members of the investigation team. The FDNY preliminary report is attached.

The U.S. Fire Administration commends FDNY for their outstanding post-mortem investigation conducted under technically difficult and politically embarrassing circumstances. The fire was tragic and the losses could have been prevented. By their investigation the FDNY has done much to help prevent similar losses in the future. The investigation also may serve as a model for other post-mortems.

SUMMARY OF KEY ISSUES

ISSUES	COMMENTS
Cause of Fire	Trash ignited in compactor chute.
Sprinkler Systems	
Compactor Chute Systems:	
13th to 35th floors	Failed to operate; heads clogged.
1st to 12th floors	Open pipe connection; inoperable.
Head at basement compactor unit	Valve closed.
Fire Department Notification	Tenants at first misinterpreted fire to be common trash fire and did not call.
Structural Problems	Compactor chute walls were 2-5/8 inch, not the 3 inch that had been approved.
	Spaces left in construction of chutes and wire chases.
	Chute hopper door missing.
Fire Inspections	Did not detect major sprinkler problems or unapproved construction features.
	Inspection results not recorded.
Fire Department Communications	Dispatchers failed to notify the fireground commander of the large number of calls received from upper floor tenants.
Incident Command System	Fireground commander failed to identify severity of situation until late; crews on roof did not communicate adequately with commander on the ground and vice versa.

INTRODUCTION

A 7 a.m. daylight fire at the Schomburg Plaza highrise apartment building in the Harlem area of New York City on March 22, 1987, took the lives of seven residents. The fire was thought to have started in trash stuck in a chute between the 27th and 29th floors. It then spread upward through the chute, an adjacent pipechase, construction openings, and ultimately through the interior walls of apartments adjacent to the chute. A misperception by the fire department that the only fire was a fire in the basement trash compactor led to a delay in rescue and extinguishment efforts on the upper floors.

The fire is described in the preliminary report (attached), which was issued by the Board of Inquiry, that the FDNY convened to investigate the history of the building, the cause and spread of the fire, and fire department operations in connection with the fire. Included with the report is the press release announcing the report, which has additional background and technical information. Note in particular the two excellent diagrams showing a top view of the garbage chute and wire chase (in the report) and a side view (in the press release).

Based on the fire department's press release, discussions with a chief on the Board of Inquiry, and other sources relating to the fire, a number of lessons can be drawn on preventing such mishaps in the future.

Almost one-quarter of the population of the United States lives in multi-family dwellings. Many of these are high-rises. There are thus tens of millions of people who may be exposed to similar problems if the lessons from such fires are not heeded.

LESSONS LEARNED

Code Enforcement

1. Those charged with inspecting life safety systems of multi-family dwellings must be trained to detect problems in the systems. Familiarity with sprinkler systems should be high on the list. Inspections of Schomburg Plaza after construction and during occupancy did not recognize that the compactor chute sprinkler system in the lower floors of the building had never been connected or tested, and the system on floors 13 through 35 had clogged heads. Also, the valve which controlled water to the sprinkler head over the compactor in the basement was not open.

2. Buildings built to code can curtail damage and provide the time to save lives. New York City has more high-rise buildings than any other city in the Nation. They have stringent codes and an excellent high-rise safety record. The Schomburg Plaza building was built to the less severe 1964 State building construction code because it was a Federally-funded project, but the requirements of even that code were not met in this building, as noted below.

3. Approved building plans are not necessarily followed in construction. The walls of the compactor chute had been approved in plans at 3 inch thickness but they were constructed 2 5/8 inches. Critical spaces were left open in constructing the pipe chase. These problems did not come to light until the fire. Inspections during construction of multiple occupancy buildings are critical to long term fire safety and must be carried out knowledgeably and diligently.

4. Records must be kept of each inspection to provide a basis for compliance and maintain a history of problems. There was no checklist completed for the latest inspection of Schomburg Plaza, which was conducted less than two months prior to the fire. This was contrary to FDNY regulations but found to be a not uncommon shortcut in paperwork.

Fireground Perceptions and Communication

5. Firefighters, as do most humans, learn to expect certain types of problems in certain situations. It is easy to fall into a mental rut and interpret information as you expect it to be rather than as it is. The buildings in Schomburg Plaza had had many minor compactor chute fires that were easily extinguished, as did many other buildings of that type. When a totally different compactor chute fire occurred – where the fire was not in the basement but high up in the chute, and breaking out into the apartments, it was still perceived to be a basement garbage fire. This perception was reinforced by an actual fire in the basement compactor at the same time. Firefighters must keep alert and cannot assume that "this one" is like the ones before.

6. Firefighters discovering hazardous conditions or fire spread should not assume that others know of it. Unless specifically aware that fire extension is known, it must be immediately reported to the company or Incident Commander. One firefighter took an elevator to a floor thought to be safe, found heavy smoke, but assumed his chief knew and did not report it. That information can be critical to saving lives.

7. The fire department needs to be absolutely certain that fire or smoke has not extended before relaxing their guard. It took 16 minutes after arrival at the scene to discover the fire on the upper floors. This was nine minutes after the rescue unit was released to return to quarters because it was thought the fire in the compactor chute was out.

Fire Dispatchers

8. Building residents made many calls to fire dispatchers only to be told that everything was being handled, when, in fact, the severity of the fire had gone unrecognized. It was assumed that the repeated calls were for the basement fire and odor of smoke. Callers were not adequately questioned as to their circumstances – did they see heavy smoke, feel heat, etc. Dispatchers must be trained to take calls seriously and not assume they know the situation, especially when there are multiple calls from different people on different floors of a building.

9. Fire dispatchers must be courteous and considerate in dealing with the public, including individuals under stress. They must take them seriously and be polite. Their conversations are recorded. A fire department's reputation can be tarnished by one ill-mannered or inadequately trained dispatcher whose handling of a call is given attention by the media, as was the case in this fire.

10. Dispatchers should help keep the fireground commander informed of special situations, such as the quantity of calls being received. The dispatch office did not notify the battalion chief on the scene of 21 telephone calls received from occupants of the 15th-33rd floors in the ten minutes before the chief decided (at 8:07 a.m.) that the fire was out and started returning companies to quarters. While the calls could have been due to smells of smoke from a minor fire, one cannot make that assumption when the risk of being wrong can lead to a disaster.

After the Fire: Department Critiques and Employee Assistance

11. Fire departments need to undertake candid, detailed critiques after a tragedy. They must not shy away even if they made errors. That is needed to avoid the same problems in the future. The New York critique in this case was extraordinarily thorough, and can serve as an example of the level of detail and candor needed to remedy problems in any good organization.

12. Fire personnel who are the target of criticism after a tragedy should be given emotional counseling and support especially where errors occurred. Reprimands or stronger actions may be needed for some, and training for others, but most will feel great anguish and may need professional help in mitigating feelings of guilt. No one wants one tragedy to lead to another. The department's morale should not be destroyed by the media and city political leaders, who can be very tough on the department in such circumstances. The leadership must bolster the department morale at such times, and put the incident in perspective while taking actions to prevent something similar from happening in the future and holding people accountable who did not perform their job well.

Public Education

13. Residents of high-rises should be alerted to their common risk from fire and asked to report suspected fires. They should not assume such fires are minor or that others have reported them. Several residents smelled smoke early in this fire but did not report it because odors of smoke from the garbage compactor were common. The first smell of smoke was subsequently found to be about 7 a.m., but the first two calls from residents to 9-1-1 came at 7:57 a.m., almost an hour later. The residents made the same error that the fire department did – assuming that only minor fires were likely.

14. Fires in the compactor chute are almost always the result of residents throwing in lit cigarettes or hot ashes. There seemed to be little attention to trying to find out who might have started the fire. It is possible that the start of the fire was accidental and totally unrecognized by the person who started it. Residents need to be taught that through such careless behavior they not only endanger their own households and lives but the property and people throughout the building. The Schomburg Plaza fire can serve as an example of the danger of such seemingly minor fires.

ATTACHMENTS

1. Fire Department of the City of New York Board of Inquiry report on the fire.

2. FDNY press release on the Board of Inquiry's findings – July 27, 1987.

3. Photographs.

BOARD OF INQUIRY
INTO THE MARCH 22, 1987
SCHOMBURG PLAZA FIRE
PRELIMINARY REPORT
June 11, 1987

INTRODUCTION

On March 22, 1987, a fire originated in the compactor chute at 1295 Fifth Avenue, Manhattan. At 07:57 hours, the fire department received a telephone call for an odor of smoke on the 29th floor. Due to an additional telephone call received from an individual complaining of smoke, a full first alarm assignment consisting of three engine companies, two ladder companies and a battalion chief was dispatched.

The building is a thirty-five story, 100 x 100 ft octagon shaped multiple dwelling of non-combustible construction. The building is part of the three building Schomburg Plaza complex which was developed by the New York State Urban Development Corporation in conformity with the 1964 New York State Construction Code. Urban Development Corporation officials were required by law to insure that the complex was built to code. As a routine matter, fire prevention inspections were also conducted by New York City Fire Department personnel.

When the first fire department units arrived at 08:00 hours, they were informed by maintenance personnel that there was a small fire in the cellar waste compactor room and that it was being extinguished.

As units continued their operations, heavily advanced fires were discovered on the 23rd, 33rd, 34th and 35th floors. As a result of the fire conditions which extended to apartments 33-H and 34-H, seven residents lost their lives.

The tragic loss of life and the unusual nature of the fire prompted Fire Commissioner Joseph E. Spinnato to convene a Board of Inquiry. The Board was formed on March 24, 1987, and was given full subpoena power. It was directed to investigate the building's construction and history, fire prevention procedures, the applicable codes, fire cause, origin, and spread, as well as operations during the fire.

The Board of Inquiry has been meeting several times a week since its inception. Under its direction, approximately 350 civilians and 50 firefighter interviews have been conducted. Each fire department unit that fought the fire was walked through the fire operation at 1295 Fifth Avenue under the Board's supervision. Communication tapes and computer printouts from the fire department, the police department, and emergency medical services have been gathered. Media videotapes filmed during the operation have been provided to the Board. A painstaking physical examination of the compactor shaft, including the removal of walls, has been conducted. The sprinkler systems have been dismantled, studied, and vouchered as evidence. Hundreds of photographs of the shaft have been taken. Physical evidence has been sent to a private laboratory and to the police laboratory for scientific tests. Architectural plans and construction records have been assembled. New York State and city records have been obtained. The relevant codes and related laws have been gathered. The Board has also subpoenaed and taken sworn testimony from many key witnesses.

In addition to digesting and analyzing those materials already gathered, much work remains to be done. However, the following general areas of concern and preliminary findings have emerged from the Board's work.

CODE

Schomburg Plaza was constructed in the early 1970s under the auspices of the New York State Urban Development Corporation (UDC). As with other UDC projects, Schomburg was built in conformity with the 1964 Building Construction Code of the State of New York, rather than the 1968 New York City Building Code.

Each code specifies a fire-resistance rating of two (2) hours for shaft enclosures in all non-combustible construction. While there is no specific section in either the State or city code dealing exclusively with compactor shaft enclosures, both codes contain general sections which indicate that shaft enclosures must be an approved assembly tested for a two-hour fire-resistance rating. The Board of Inquiry has determined that the "as built" plans for Schomburg Plaza specify three (3) inch enclosure walls for the compactor shaft. Such a design, if properly constructed, would comply with both State and city codes.

However, examination of the shaft discloses that it was not built according to plan. The wall assembly was two and five-eights inches (2 5/8 inch) thick, not the three inches (3 inch) called for in the "as built" plans. The Board has been unable to find any test which indicates that a two and five eights inch (2 5/8 inch) assembly has a two-hour fire-resistance rating. The Board is continuing its analysis of code requirements and approved assemblies as they relate to shaft enclosures and compactors and will make further comments in its final report.

Both the city and State codes require sprinklers inside the compactor chute.

FEBRUARY 3, 1987 INSPECTION

On February 3, 1987, Engine Company 91, conducted a regularly-scheduled yearly inspection of the three buildings of Schomburg Plaza. The inspection was conducted by a covering lieutenant and five firefighters from Engine Company 91. The inspection took approximately 40 minutes.

Regulations require that a checklist be completed for each building. The checklist notes types of items that the firefighters are to inspect. The checklist contains a section dealing with sprinklers and their associated control valves (OS and Y valves) but does not specifically deal with sprinklers inside compactor chutes (see attached blank checklist). No checklist was completed for any of the three buildings during the February 3, 1987, inspection.

The Board of Inquiry has determined that, at the time of the fire, the compactor sprinkler system did not work and that neither of the two OS and Y valves which controlled the sprinkler system were open at the time of the inspection. While one of the firefighters believed that he bad inspected an OS and Y valve for the compactor sprinkler system at 1295 Fifth Avenue, he had mistaken an open OS and Y valve which controlled the supply of domestic fresh water to the building for the sprinkler valve. The valve was located in the basement next to the compactor.

The Board of Inquiry has also determined that, as a matter of practice, required checklists for these types of inspections were rarely completed prior to the Schomburg Plaza fire. Corrective actions have been implemented by the fire department to ensure that the checklists are completed. The Board will complete an analysis of the department's inspectional procedures and may make further recommendations to the Commissioner.

FIRE ORIGIN

Based on the fire scene examination and subsequent investigation, it was determined that the fire originated in combustible material (household rubbish) within the confines of the metal compactor rubbish chute as the result of an obstruction within the compactor chute in an area between the 27th and the 29th floors. The manner of ignition is unknown but the most likely cause was the careless disposal of something similar to a lit cigarette.

Because the compactor chute sprinkler system was not working, the original fire was not extinguished and extended vertically within the compactor chute. This vertical extension resulted from the burning of the original combustible material and the ignition of the flammable residue which had accumulated on the interior surface of the chute. This unchecked fire condition caused a high-heat buildup within the compactor chute at the upper floors.

The examination and investigation indicated that the fire further extended as follows:

1. To and throughout the compactor closet on the 29th floor via the missing chute hopper door and into the public hallway on the 29th floor via the open compactor closet door;

2. Into the kitchen areas of apartments 23-H, 33-H, and 34-H via convection and radiation transferred through construction openings, and via conduction through various metal pipe support brackets and chute anchors, and thereafter extended to and throughout the above apartments via flame spread; and,

3. Into apartment 35-H by auto-exposure via the windows from apartment 34-H.

The extension of the fire into apartments 33-H and 34-H was the proximate cause of the seven fire fatalities.

COMPACTOR CHUTE AND COMPACTOR SPRINKLER

The compactor chute investigation disclosed that the heat buildup from the fire within the chute was transferred to the metal chute itself. The heated metal chute radiated heat into the drywall chute enclosure and the adjoining pipe chase voids, which are adjacent to the "H" bank apartments. The heat buildup in the shaft enclosure compartment traveled horizontally into the adjoining pipe chase void via a one and one-eighths inch (1-1/8") opening at the bottom of the partition wall between the shaft enclosure and the pipe chase void. This opening was created as a result of the two inch (2") drywall wall resting on top of various pipe brackets and chute anchors rather than being secured to the concrete floor. (See attached diagram.) In addition, it was found that the chute penetrated each floor through a concrete opening of a slightly larger diameter; thereby, creating a space between the outside of the chute and the concrete opening which varied from floor to floor. These openings allowed for vertical heat transfer from floor to floor within the shaft. Examination found additional openings in the drywall construction of the walls which divided the compactor chute enclosure and the pipe chase void from the kitchen and living rooms of the "H" bank apartments. These openings allowed heat transfer from the chute enclosure and the pipe chase void into the "H" bank apartments. Further, the above mentioned metal pipe brackets extended from the pipe chase void into and under the kitchen broom closet, making contact with the combustible components of the broom closet.

The compactor chute sprinkler investigation disclosed that the system was comprised of two separate and distinct systems. The lower system was designed to extinguish fires below the 13th floor. The upper system was designed to extinguish fires occurring between the 13th and 35th floors.

Examination of the lower system revealed an open pipe connection between the sprinkler head and the related piping. This opening in the system proved that the lower sprinkler system was not operable at the time of the fire.

Investigation of the upper sprinkler system during the initial fire investigation revealed that sprinkler heads were clogged with rust and silt at various floors. Additionally, the evaluation of the extensive fire damage on the upper floors indicated that the upper sprinkler system failed to operate at the time of the fire.

Examination of the single sprinkler head located at the compactor unit in the basement revealed that this head was controlled by a shut-off valve which additionally controlled a cold water hoseline. Investigation revealed an absence of water flow at this hoseline and sprinkler head at the time of the fire, indicating that the valve was closed.

The Board has concluded that the original fire started in combustible rubbish confined within the compactor chute which intensified due to the lack of a properly operating sprinkler system. The fire was able to connect and extend into various apartments due to the previously described metal brackets, opening, and voids.

FIRE OPERATION

The first indication of the fire was an odor of smoke which several residents have since reported smelling sometime after 7 a.m. Initially, no one was concerned because odors of smoke from compactor fires were not uncommon at 1295 Fifth Avenue.

Maintenance employees investigated, found, and proceeded to extinguish with a garden hose, a fire burning in the compactor in the basement. At the same time, security guards employed at Schomburg Plaza investigated internal reports of smoke on the upper floors. It was not until 07:57 hours that two phone calls were received by 9-1-1. The first call was from a tenant in apartment 29-H complaining of smoke in the hallway, the second from a security guard reporting smoke on the twenty-fourth floor.

At 07:57 hours, because of these phone calls, a full first alarm assignment of three engines and two ladders was dispatched. Automatically, a computer-generated message was received by these companies via teletype and by the fire department dispatchers on computer screens. The message specified that 1295 Fifth Avenue was a multiple dwelling of 34 stories measuring 100 feet by 100 feet and further specified "compactor fires may require additional ladder company for severe smoke condition on upper floors."

On arrival in the lobby at 08:00, after a three minute response time, the fire department was informed by maintenance personnel that the fire was in the compactor room and was being extinguished. A hoseline was stretched to the compactor chute on the first floor. An examination of the compactor in the basement disclosed a rubbish fire being extinguished by maintenance personnel. At the same time, two firefighters were sent to the roof to perform ventilation work and to begin a survey of the upper floors.

At 08:07, the battalion chief's aide transmitted a radio message indicating a compactor fire in the basement. He stated that the fire had been extinguished and he was using one engine company and two ladder companies due to a heavy smoke condition on the upper floors. Because it was believed the fire was out the rescue company was returned. Two engine companies stood by. The battalion chief was not told and was unaware that between 07:57 and 08:07 hours the Dispatch Office had received more than twenty-one telephone calls from tenants on the fifteenth through the thirty-third floors reporting large amounts of smoke in hallways and apartments. Some of these calls were from occupants describing extremely heavy smoke conditions.

The two firefighters arrived on the roof at approximately 08:06 and began venting and examining the roof. A second team consisting of two firefighters arrived on the roof at approximately 08:10 and also engaged in examination and venting work.

At approximately 08:11 one of the four roof men descended to the upper floors. At approximately 08:16 this roof man discovered the fire in apartment 34-H and transmitted an urgent report of the fire. Units immediately began taking hoselines to the upper floors to extinguish the fire. A few seconds later, three members of the Jenkins' family jumped from apartment 33-H. The firefighters reacted quickly and began applying water to the fire in apartment 33-H at 08:22 hours. Operations were initially hampered by low water pressure. During the course of extinguishment and control, four additional victims were discovered in apartments 33-H and 34-H. Additional units were dispatched to the scene at 08:17, 08:21, and 08:25 hours. At 08:35, a second alarm was transmitted. The fire was declared under control at 09:45 hours.

The Board is concerned with the approximately sixteen (16) minutes between the arrival of the first unit and the discovery of the fire on the upper floors. It was made a preliminary determination that because of the conditions visible on arrival and the belief that this was a routine compactor fire, similar to many previous compactor fires at Schomburg Plaza, neither the firefighters nor the dispatchers recognized significant information indicating that this was not a normal compactor fire. The Board is also concerned with some operational decisions made after 08:16 hours, but it is clear that by 08:16 hours nothing could have been done to limit the fire fatalities.

Div.	Batt.	Co.	Block No.

MULTIPLE DWELLING INSPECTION FORM   A-291 (10/86) 25-860384-R040

Address _____

Type Building (Check One)   MD "A" _____   NLT _____   OLT _____   CD "A" _____

Inspector: Fill in all blanks. See Instruction Form A-291A

A. ROOF

1. Clear of Rubbish _____ (FP-100)
2. Gooseneck ladder secure _____ (FP-103)
3. Incinerator stack equipped with approved spark arrester _____ (A8-41) arrester clean and in good repair _____ (SP-8)
4. Gravity tank free of leaks _____ (SP-43-49) supports free of rust _____ (SP-44-50) structural defects _____ (A8-34)
5. Roof brickwork and masonry in good repair _____ (A8-9)
6. TV antennae, wires, 10 feet or more above roof _____ (SP-70)
7. Bulkhead or scuttle opening within 4' of roof edge has guard rail or parapet protection _____ (A8) in good condition _____ (SP)

B. HALLWAY(S) and STAIRS

1. Clear of rubbish _____ (FP-100) Location _____
2. Unobstructed _____ (FP-28-29) Location _____
3. Approved means of egress (inadequate means, consult officer).
   a. MD "A", NLT, OLT more than 2 stories has 2 means of egress _____ (A8-8)
   b. CD "A" more than a basement and 2 stories has 2 means of egress or 1 means and sprinklered halls and stairs _____ (A8-8)
      when sprinklered, report of annual sprinkler test is current _____ (SP-76)
4. Scuttle cover openable _____ (FP-105, FP-106) fixed iron ladder _____ (SP-11, A8-43) unenclosed scuttle ladder _____ (A8)
   bulkhead door openable _____ (FP-105, FP-106) sliding bolts or hooks or approved type knob or panic bolt used _____ (FP-105)
   (See instructions)
5. Apartment doors to public hallways.
   a. In MD "A" and NLT are Fireproof _____ (A8-40) self-closing _____ (SP-17) good operating condition _____ (SP-18) main-
      tained closed _____ (SP-18) Location _____
      Following apartment doors have been checked _____
   b. In OLT and CD "A" are self-closing _____ (SP-17) good operating condition _____ (SP-18) maintained closed _____ (SP-18)
      any glazing must be wired glass _____ (SP-25) Location _____
      Following apartment doors have been checked _____
   c. OLT 4 stories or more have fireproof doors _____ (A8-40) (See instructions)
      Location _____
      Following apartment doors have been checked _____
   d. Free of supplementary louvred or screen doors _____ (SP-72) Location _____
6. a. Where stairways are enclosed, doors are self-closing _____ (SP-17) good operating condition _____ (SP-18) maintained closed
      _____ (SP-18) Location _____
   b. Stairway lighting maintained _____ (SP-19)
7. Elevator shafts are enclosed _____ (A8-7) Fireproof doors _____ (A8-40) Self-closing _____ (SP-17) maintained closed
   _____ (SP-18) Location _____
8. Floor signs present _____ (A-8) Location _____
9. Transoms in OLT and CD "A" are fixed shut and any glazed opening in door transom are of wire glass
   Location _____
   Other transom type openings in public halls are replaced with fire retarding materials _____ (SP-24, 25, A8-51, 54)
   Location _____
10. Dumbwaiter doors are fireproof _____ (A8-40) self-closing _____ (SP-15) good operating condition _____ (SP-16) main-
    tained closed _____ (SP-16) unused dumbwaiter shafts are sealed off _____ (A8-38, SP-81)
    Location _____
    Other unused shafts or ducts sealed off _____ (A-38) Location _____
    Incinerator doors and hoppers are fireproof _____ (A8-40) self-closing _____ (SP-15) good operating condition _____ (SP-16)
    maintained closed _____ (SP-16) Location _____
11. Cellar Stairs
    a. In MD "A" and NLT stairs within bldg. are equipped at top and bottom with fireproof doors _____ (A8-40) self-closing _____
       (SP-15) good operating condition _____ (SP-16) maintained closed _____ (SP-16)
    b. In OLT (more than a basement and 3 stories) door at top is fireproof _____ (A8-40) self-closing _____ (SP-15) good operating
       condition _____ (SP-16) maintained closed _____ (SP-16)
    c. In CD "A" stairs are enclosed and door at top or at bottom is fireproof _____ (A8-40) self-closing _____ (SP-15) good operating
       condition _____ (SP-16) maintained closed _____ (SP-16)
12. Standpipe hose and outlets in good condition _____ (FP-42) Location _____

1295 5TH AVE MAN. 33RD FLOOR COMPACTOR CHUTE AREA *

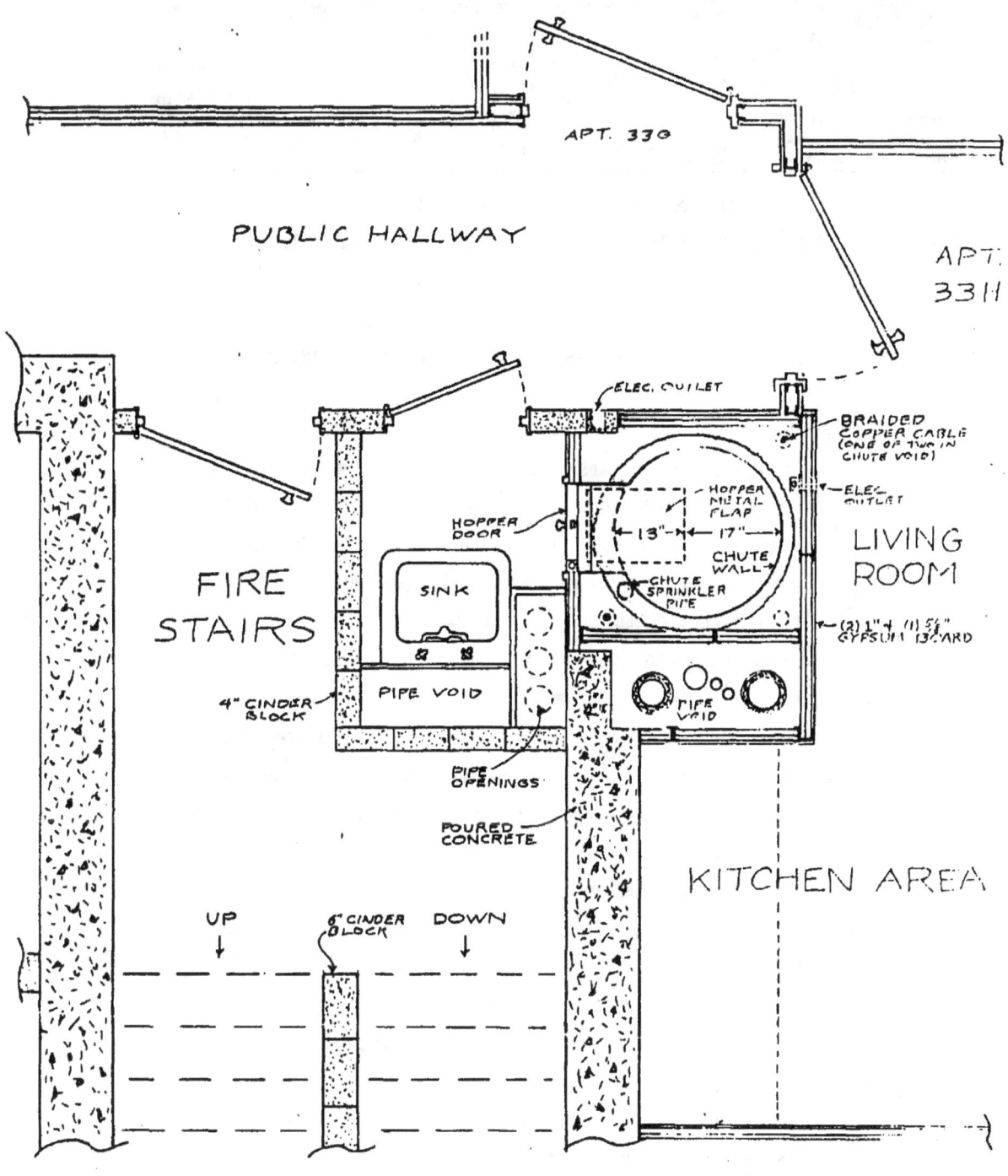

* SOURCE: ON-SITE MEASUREMENTS

Prepared By The
GRAPHICS UNIT
F.D.N.Y.

Fr. Mulcahy 4/21/87

FIRE DEPARTMENT NEW YORK

250 LIVINGSTON STREET • BROOKLYN, N.Y. 11201-5884
100-87

JOSEPH E. SPINNATO, Fire Commissioner

OFFICIAL.
NEWS

1:00 P.M. Monday

For Release

July 27, 1987

For Further Information. Call 403-1543-44
John Mulligan Assistant Commissioner
Lt. Frank Martinez Press Secretary and Efrain Parrilla

REMARKS BY FIRE COMMISSIONER JOSEPH E. SPINNATO

PRESS CONFERENCE ON BOARD OF INJURY PRELIMINARY REPORT

ON SCHOMBURGE PLAZA FIRE

MONDAY, JULY 27, 1987 - 1:00 P.M.

FIRE DEPARTMENT HEADQUARTERS, 250 LIVINGSTON ST., BROOKLYN

GOOD AFTERNOON.

WE HAVE CALLED THIS PRESS CONFERENCE TO RELEASE THE PRELIMINARY REPORT OF THE SPECIAL BOARD OF INQUIRY I APPOINTED TO LOOK INTO THE CAUSES OF THE TRAGIC MARCH 22 FIRE AT THE SCHOMBURG PLAZA APARTMENT COMPLEX.

SEVEN LIVES WERE LOST IN THAT FIRE. AS YOU MAY RECALL, AT THE TIME PEOPLE WERE ASKING HOW COULD SUCH A TRAGEDY OCCUR IN OUT CITY, GIVEN ALL OUT STRINGENT FIRE SAFETY REGULATIONS, OUR DEMANDING BUILDING CODES AND OUT EXTENSIVE FIRE FIGHTING RESOURCES.

IN APPOINTING THE BOARD OF INQUIRY, I WAS DETERMINED TO GET TO THE BOTTOMOF THIS CASE, TO DETERMINE THE CAUSES OF THE FIRE, TO SEEK TO FIND WHERE THE FAULT LAY FOR WHAT HAPPENED AND THE DEVELOP MEASURES THAT WOULD HELP US AVOID A REPETITTION OF THE TRAGEDY.

AS YOU READ THIS PRELIMINARY REPORT, YOU WILL CLEARLY SEE THAT THE BOARD HAS CONDUCTED ITS INVESTIGATION WITH COMPLETE INDEPENDENCE AND OBJECTIVITY. IT HAS SPARED NO ONE.

BEFORE GOING INTO THE SUBSTANCE OF THE REPORT, HOWEVER, I WANT TO CLARIFY A CRITICAL POINT CONCERNING A STORY WHICH APPEARED IN YESTERDAY'S PRESS.

IT WAS PARTICULARLY DISTRESSING TO READ A HEADLINE WHICH STATED, 'DISPATCHERS LAUGHED AT FIRE-VICTIMS' CALLS.'

-MORE-

-COMMISSIONER SPINNATO'S REMARKS- PAGE 2

I WANT TO EMPHASIZE THAT THIS HEADLINE, AS WELL AS THE ONE INSIDE TO THE EFFECT THAT 'DISPATCHERS LAUGHED OFF HARLEM FIRE,' GROSSLY MISREPRESENT THE FACTS. THERE IS NO EVIDENCE WHATSOEVER THAT DISPATCHERS LAUGHED AT THE CALLERS FROM SCHOMBURG.

THE BOARD OF INJURY DID FIND INSTANCES OF INSENSITIVE TREATMENT OF CALLERS FROM SCHOMBURG AMONG THE MANHATTAN DISPATCHERS. SUCH UNPROFESSIONAL BEHAVIOR IS INTOLERABLE TO THE ADMINISTRATION OF THE FIRE DEPARTMENT, AND WE HAVE TAKEN ADMINISTRATIVE ACTION AGAINST THOSE SPECIFIC DISPATCHERS AND ARE CONSIDERING FURTHER DISCIPLINARY MEASURES AGAINST THEM.

BUT AT NO TIME IN THE RECORDING FO THE DISPATCHERS" EXCHANGES WITH THESE CALLERS IS THERE ANY EVIDENCE OF LAUGHTER.

NONETHELESS, I AM DEEPLY DISTURBED BY THE EVIDENCE OF INSENSITIVE TREATMENT OF CALLERS, AS WELL AS BY FINDINGS THAT POINT OUT OTHER SHORTCOMINGS WITHIN THE DEPARTMENT.

I AM DETERMINED THAT NO ONE WHO IS FOUND TO BE AT FAULT IN THIS TRAGIC INCIDENT GOES UNPUNISHED. WE DEMAND THE HIGHEST LEVEL OF PROFESSIONALISM FROM OUR MEMBERS AND INDEED OUR PEOPLE PERFORM IN THAT MANNER EACH AND EVERY DAY. HOWEVER, IF AND WHEN WE DO FALL BELOW THAT LEVEL OF EXPECTATION WE MUST AND DO TAKE CORRECTIVE ACTION.

IT IS EXTREMELY IMPORTANT TO POINT OUT THAT THE BOARD HAS FOUND SERIOUS SHORTCOMINGS IN THE CONSTRUCTION OF THE SCHOMBURG PLAZA AND IN THE MAINTENANCE OF THE BUILDING'S COMPACTOR SPRINKLER SYSTEM, IT IS IMPORTANT TO UNDERSTAND THAT THESE CONSTRUCTION AND MAINTENANCE SHORTCOMINGS WERE THE PROXIMATE CAUSE OF THE SEVEN DEATHS THAT FATEFUL SUNDAY MORNING. BOARD CHAIRMAN JONATHAN FAIRBANKS WILL GO INTO THESE AND OTHER DETAILS OF THE REPORT IN A MINUTE.

AS I HAVE SAID, THE BOARD ALSO FOUND PROBLEMS WITH THE DEPARTMENT'S OPERATIONS AND PROCEDURES. THESE PROBLEMS COVER AREAS SUCH AS THE TIMELY INSPECTION OF BUILDINGS , THE PERFORMANCE OF FIRE ALARM DISPATCERS AT THE TIME OF THE SCHOMBURG PLAzA FIRE AND THE FIREGROUND OPERATIONS AT SCHOMBURG INCLUDING THE MANAGEMENT OF THE FIRE DURING THE EARLY STAGES OF THE INCIDENT.

(MORE)

REMARKS BY FIRE COMMISSIONER SPINNATO - PAGE 3

WHILE I AM TROUBLE BY THE FINDINGS IN THE REPORT, I AM PROUD OF THE WORK DONE BY THE BOARD OF INQUIRY. IT HAS ENABLE THE FIRE DEPARTMENT TAKE A HARD AND DETAILED LOOK AT ITSELF AND WHERE NECESSARY TAKE CORRECTIVE ACTIONS.

WE HAVE DEVELOPED A WIDE-RANGING SET OF MEASURES TO CORRECT OPERATIION- DIFFICULTIES UNCOVERED BY THE BOARD.

SOME OF THESE, INCLUDING A MORE AGGRESSIVE PROGRAM OF BUILDIG INSPCETIONS, ARE ALEADY BE PUT IN PLACE, WHILE OTHERS WILL SHORTLY BE IMPLEMENTED.

SOME REQUIRE THE COOPERATION OF THE FIRE UNIONS AND WE ARE HOPEFUL THE UNIONS WILL JOIN THE ADMINISTATION OF THE DEPARTMENT IN THIS EFFORT.

MEANWHILE, THIS PRELIMINARY REPORT HAS BEEN FORWARDED TO THE MANHATTAN DISTICT ATTORNEY FOR PURSUIT OF ANY AND ALL POSSIBLE CRIMINAL VIOLATIONS DISCLOSED BY OUR INVESTIGATION.

AND AS I HAVE SAID, ANY DEPARTMENT PERSONNEL WHO ACTED NEGLIGENTLY OR IN VIOLATION OF THIS DEPARTMENT'S RULES OR REGULATIONS WILL BE HELD ACCOUNTABLE THE DEPARTMENT'S DISCIPLINARY PROCEDURES.

NOW I WOULD LIKE BOARD CHAIRMAN FAIRBANKS TO COME THE MICROPHONE AND FILL YOU ON THE DETAILS, AFTER WHICH WE WILL TAKE YOUR QUESTIONS.

-30-

PHOTOGRAPHS FROM
THE FIRE DEPARTMENT OF THE CITY OF NEW YORK
BUREAU OF FIRE INVESTIGATIONS

The photos listed below are representative of construction and maintenance practices throughout the entire Schomburg Plaza building.

1. Exterior of fire building with water and burn patterns.

2. Compactor unit in the fire building (morning of fire).

3. Kitchen area, Apartment 23, adjoining compactor and pipe chase.

4. Overall view of relationship between kitchen area and pipe void and compactor chute.

5. Heat buildup in compactor closet.

6. Typical compactor closet, from public hall.

7. Opened wall in compactor closet to view chute interior.

8. Visual inspection. (Typical of all floors)

9. Opening of apartment walls for interior examinations.

10. Sprinkler system in disrepair. Located at top of chute hopper.

11. Pipe bracket extending into kitchen from pipe void compromising firewall between void and living area.

12. Void between chute and structure floor allowing for passage of heat from floor to floor.

13. Interior of chute with opening at seam.

14. Clogged sprinkler head.

15. Horizontal piping with solidified silt and rust.

16. Evidence voucher. (Sample)

17. Sprinkler shut off valve in dayroom, first floor.

18. Sprinkler shut off valve in closet (compactor) on 35th floor at ceiling.

Photo 1

Photo 2

Photo 3

Photo 4

Photo 5

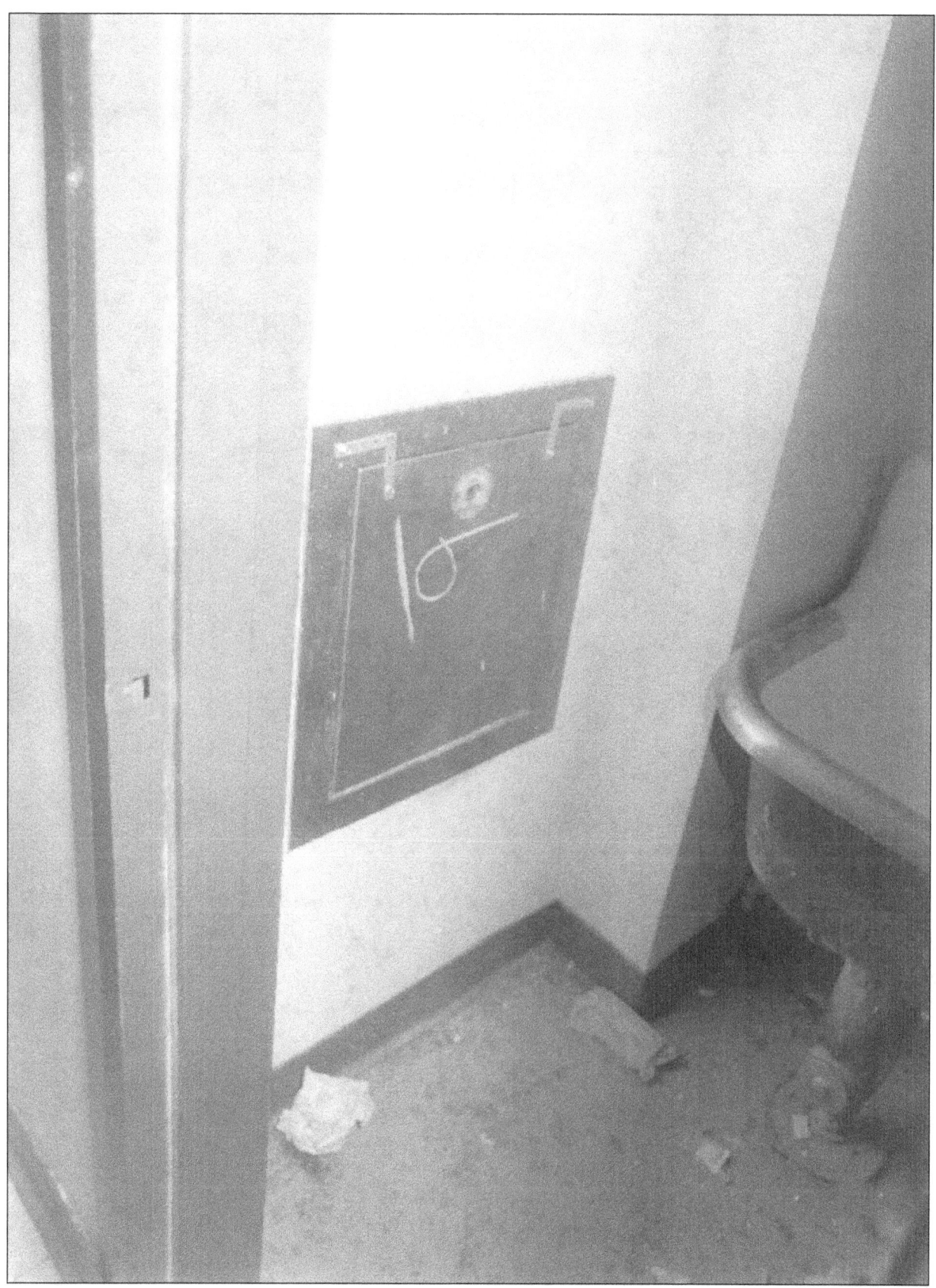

Photo 6

Photo 7

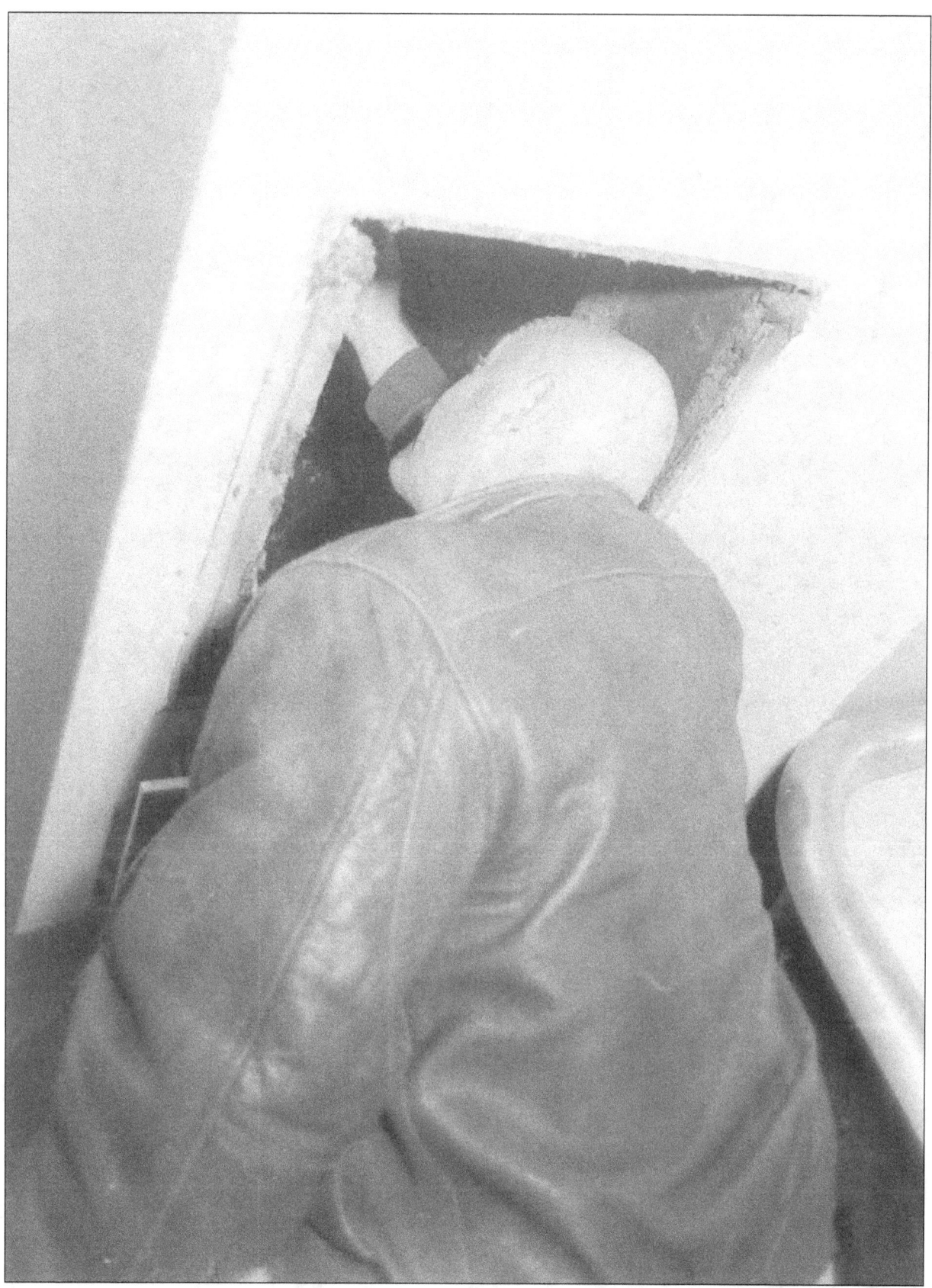

Photo 8

Photo 9

Photo 10

Photo 11

Photo 12

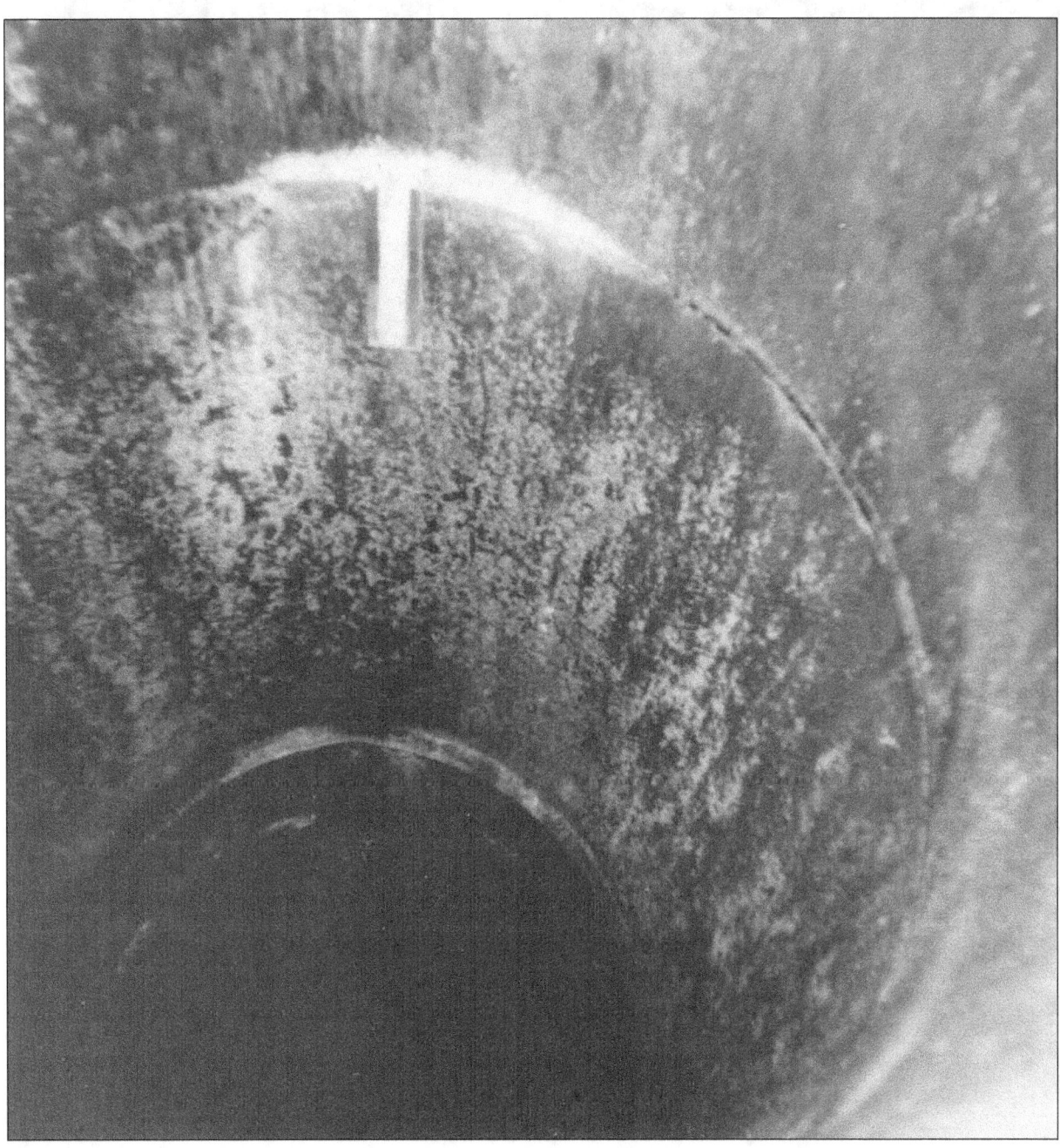

Photo 13

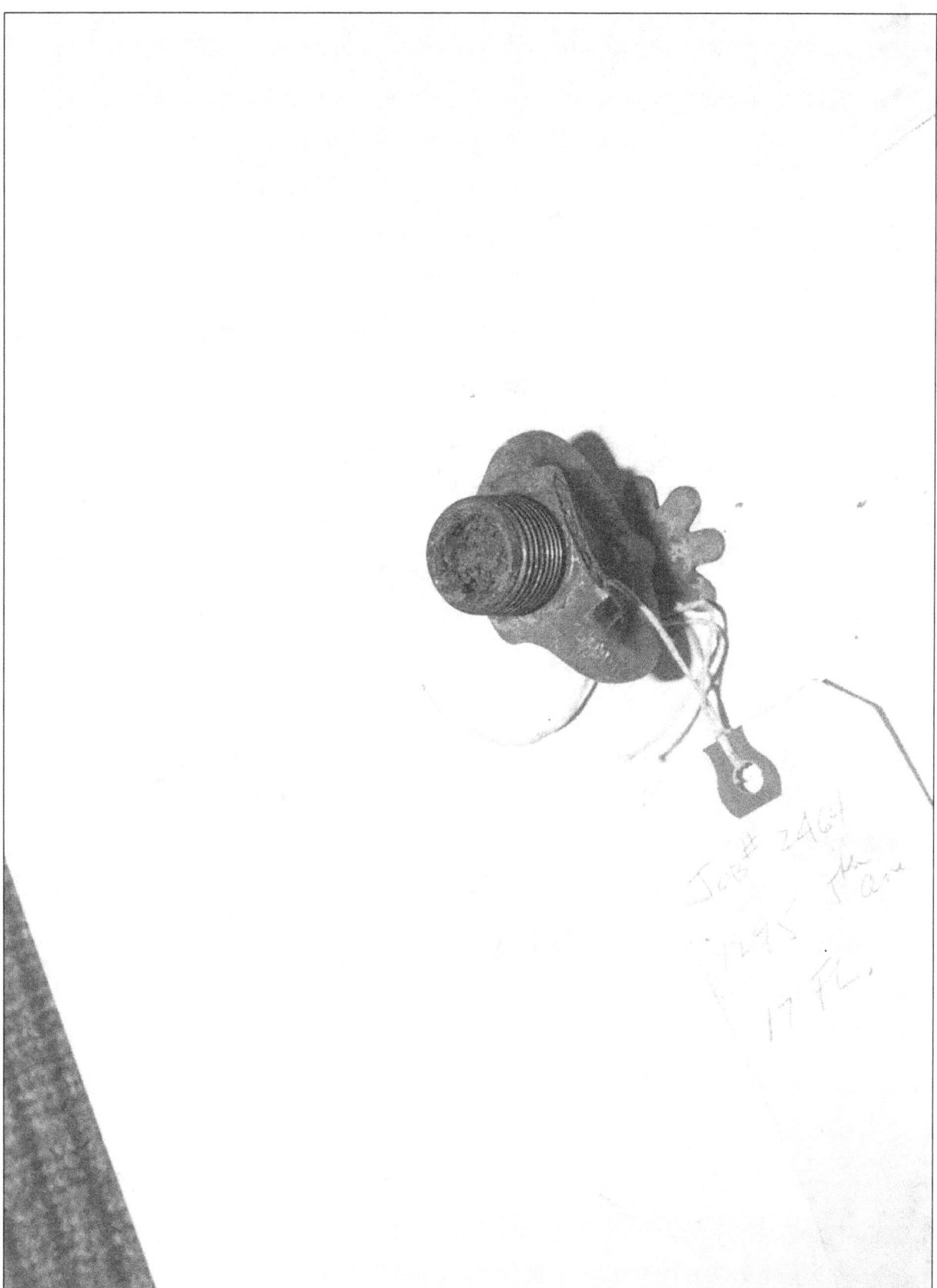

Photo 14

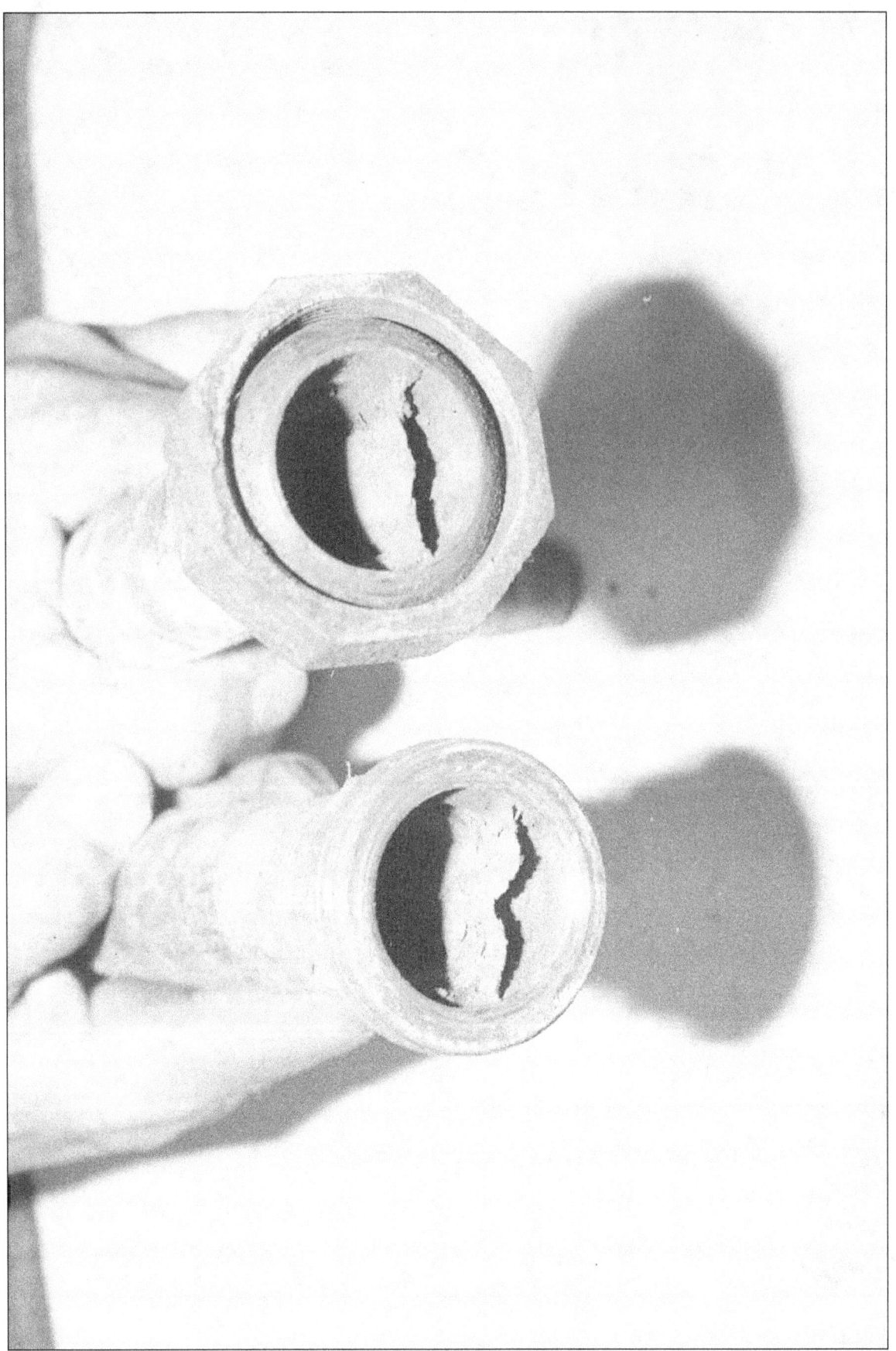

Photo 15

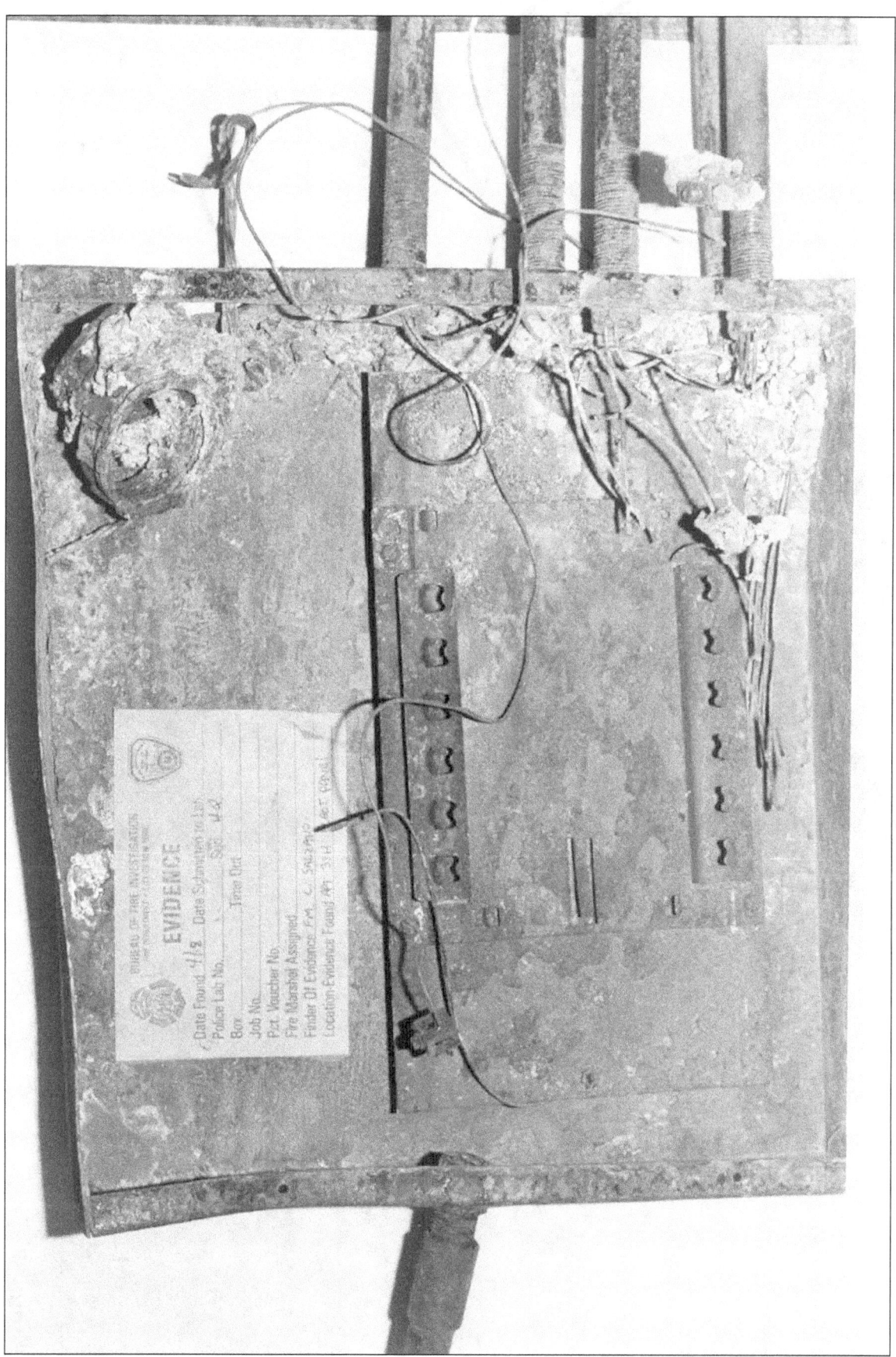

Photo 16

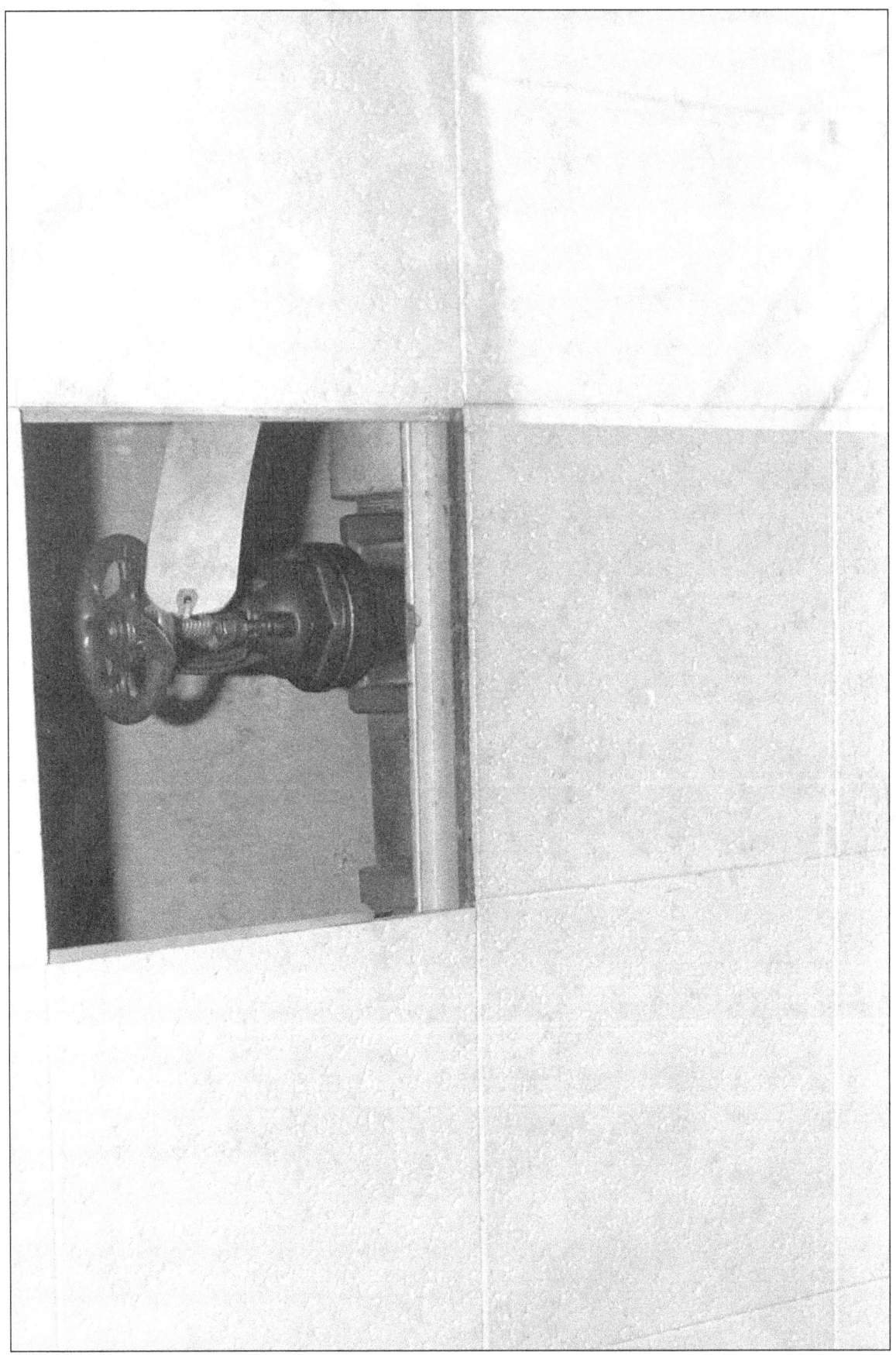

Photo 17

Photo 18